CHEVY TRUCKS

Publications International, Ltd.

Copyright © 2019 Publications International, Ltd.
All rights reserved. This book may not be reproduced
or quoted in whole or in part by any means whatsoever
without written permission from:

Louis Weber, CEO
Publications International, Ltd.
8140 Lehigh Avenue
Morton Grove, IL 60053

Permission is never granted for commercial purposes.

ISBN: 978-1-64030-666-0

Manufactured in China.

8 7 6 5 4 3 2 1

PHOTOGRAPHY:

The editors would like to thank the following people and organizations for supplying the photography that made this book possible. They are listed below, along with the page number(s) of their photos.

Roland Flessner: 128, 129; **General Motors Company:** 24, 25, 26, 27, 28, 29, 32, 33, 42, 43, 84, 85, 112, 113, 118, 119, 130, 131, 132, 133, 134, 135, 136, 137, 138, 139, 140, 141, 142, 143, 144; **Thomas Glatch:** 44; **Conrad Gloos:** 102, 103, 104, 105; **Sam Griffith:** 15, 16, 17, 54, 55, 56, 57; **Dan Lyons:** 70, 71; **Vince Manocchi:** 8, 9, 22, 23, 37, 38, 39, 40, 41, 46, 47, 50, 51, 60, 61, 74, 75, 94, 95, 124, 125; **Roger Mattingly:** 12, 13, 14; **Doug Mitchel:** 18, 19, 20, 21, 30, 31, 34, 35, 58, 59, 63, 72, 73, 76, 77, 78, 79, 96, 97, 106; **Mike Mueller:** 6, 7; **Nina Padgett:** 10, 11; **Blake Ramick:** 108, 109; **Al Rogers:** 64, 65, 66, 67; **Tom Shaw:** 92, 93; **Steve Statham:** 90, 91; **David Temple:** 86, 87, 114, 115; **Bob Tenney:** 98, 99; **Phil Toy:** 68, 69, 80, 81, 82, 83, 116, 117; **W.C. Waymack:** 45, 48, 49, 88, 89, 100, 101, 107, 120, 121

ADDITIONAL ART:

SHUTTERSTOCK.COM

Our appreciation to the historical archives and media services groups at General Motors Company.

OWNERS:

Special thanks to the owners of the trucks featured in this book for their cooperation. Their names and the page number(s) for their vehicles follow.

Nelson Bates: 114, 115; **Biff Behr:** 12, 13, 14; **Dan Brown:** 90, 91; **Mark Bryant:** 116, 117; **Bill Cotherman:** 30, 31; **Richard DeVecchi:** 34, 35; **Ernie's Wrecker Service:** 56, 57; **Fairway Chevrolet:** 10, 11; **Bob H. Firth:** 98, 99; **General Motors Company:** 24, 25, 26, 27, 28, 29, 32, 33, 42, 43, 84, 85, 112, 113, 118, 119, 128, 129, 130, 131, 132, 133, 134, 135, 136, 137, 138, 139, 140, 141, 142, 143, 144; **Whitney and Diane Haist:** 68, 69; **William T. Hayes and sons:** 54, 55; **Bob Howard:** 108, 109; **Chesley Jacobs:** 62, 63; **Bob Kamerer:** 102, 103, 104, 105; **Bill Kaprelian:** 58, 59; **Terry Knight:** 48, 49; **Bill and Diann Kohley:** 63; **Bob Lanwermeyer:** 88, 89; **Norbert C. Laubach:** 96, 97; **Ricky Lorenzen:** 94, 95; **Raymond L. May:** 72, 73; **Peter Mellinger:** 92, 93; **David Meyer:** 76, 77, 78, 79; **Danny Naile:** 107; **Herman Pfauter:** 37, 38, 39, 40, 41; **Scott Pickle:** 22, 23; **Steve Provart:** 100, 101; **Gary Romoser:** 45; **Lydia and Byron Ruetten:** 106; **Eldon Schmidt:** 50, 51; **William Schoenbeck:** 15, 16, 17; **Jerry Shumate:** 120, 121; **Tom Slusser:** 44; **Tom Snively:** 6, 7; **Larry and Susan Steemke:** 124, 125; **Dennis Syphrett:** 86, 87; **Michael Thomas:** 81, 81, 82, 83; **William and Patricia Thomas:** 70, 71; **Robert Warnick:** 46, 47; **Gordon Watson:** 64, 65, 66, 67; **Ron Willemsen:** 74, 75; **Don Wolf:** 18, 19, 20, 21; **Daniel Wright:** 8, 9

CONTENTS

1918 490 LIGHT DELIVERY **6**
1919 490 LIGHT DELIVERY **8**
1925 SUPERIOR K **10**
1928 LIGHT DELIVERY WAGON **12**
1928 PICKUP **15**
1929 DUMP TRUCK **18**
1930 PICKUP **22**
1933 SEDAN DELIVERY **24**
1936 CANOPY DELIVERY **26**
1936 PICKUP **30**
1936 SUBURBAN CARRYALL **32**
1941 PICKUP **34**
1942 G-7117 MILITARY **37**
1946 SUBURBAN **42**
1946 PICKUP **44**
1946 PANEL **46**
1947 PANEL **48**
1948 CANOPY EXPRESS **50**
1949 PICKUP **54**
1949 TOW TRUCK **56**
1950 SEDAN DELIVERY **58**
1952 PICKUP **60**
1954 PICKUP **62**
1955 SEDAN DELIVERY **64**
1955 PANEL **68**
1955 PICKUP **70**
1956 PICKUP **72**
1958 PICKUP **74**
1958 PICKUP **76**
1959 PICKUP **80**

1959 EL CAMINO	**84**
1960 PICKUP	**86**
1961 CORVAIR RAMPSIDE	**88**
1961 PICKUP	**90**
1961 PANEL	**92**
1962 PANEL	**94**
1963 CORVAIR RAMPSIDE	**96**
1965 EL CAMINO	**98**
1965 PICKUP	**100**
1966 EL CAMINO	**102**
1966 PICKUP	**106**
1968 PICKUP	**107**
1970 PICKUP	**108**
1970 EL CAMINO	**110**
1971 PICKUP	**112**
1972 PICKUP	**114**
1972 BLAZER	**116**
1975 PICKUP	**118**
1977 EL CAMINO	**120**
1982 S-10	**122**
1983 EL CAMINO	**124**
1988 PICKUP	**126**
1990 PICKUP	**128**
2002 AVALANCHE	**131**
2003 SSR	**132**
2004 COLORADO	**134**
2007 PICKUP	**135**
2014 PICKUP	**136**
2015 COLORADO	**138**
2018 COLORADO ZR2	**140**
2019 PICKUP	**142**

1918 490 LIGHT DELIVERY

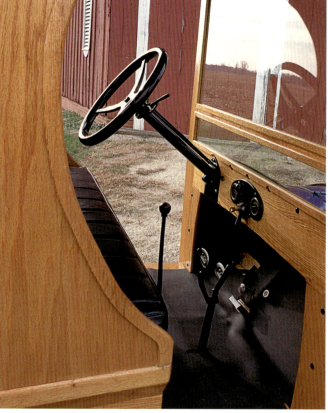

Chevrolet's first commercial vehicles debuted for 1918, the same year the company officially became a part of General Motors. Like other manufacturers of the day, Chevrolet built only the complete chassis and cowl, leaving the bodywork to various commercial coachbuilding firms. The Model 490 Light Delivery chassis sold for $595—body not included. This 490 pickup illustrates the ratio of Chevrolet/outside supplier components clearly—the wooden portion of the truck is the part Chevy didn't furnish. The Light Delivery chassis was mostly shared with Chevrolet's passenger cars but was fitted with heavier springs that enabled it to handle a payload of 1000 pounds. The engine—a 171 cubic-inch four-cylinder rated at 26 horsepower—was also shared with the 490 passenger-car lineup. Chevrolet's first one-ton truck, ironically dubbed Model T, also joined the roster this year. It used modified mechanicals from Chevy's upscale Model FA passenger car.

1918 **490 LIGHT DELIVERY**

1919 490 LIGHT DELIVERY

The popular 490 series of vehicles was little changed for 1919, though the Light Delivery got an electric starter. This restored beauty is equipped with an accessory water-temperature gauge atop the radiator, a period device generally known as a "Moto-Meter." (Moto-Meters bearing the Chevrolet trademark weren't offered until 1923, however.) The cargo area featured a drop-down tailgate, along with curtains for semi-secure hauling. Note also the roof-mounted turn signals, caged dome light, and the hand-operated horn nestled next to the driver's seat behind the spare tire. As ever, the 490's little four-cylinder engine wasn't much for speed; rugged reliability and easy maintenance counted for a lot more in those days, and these Chevrolets delivered.

1919 490 LIGHT DELIVERY

1925 SUPERIOR K

10 | **CHEVY TRUCKS**

Chevrolet introduced its Superior Series K models in January 1925. Among the many mechanical improvements was a completely revised engine with a newly designed block, heavier crankshaft, and drop-forged connecting rods with larger bearings. This Series K Light Delivery has been fitted with a "C-cab" panel body, a style that was becoming a bit outdated by the mid Twenties. Instrumentation on all Series K Superiors was small and centrally grouped. As the year wore on, spark and throttle controls in the cars were moved to the dash, but trucks retained these levers on the steering column.

1928 LIGHT DELIVERY WAGON

1928 LIGHT DELIVERY WAGON | 13

In the automotive world, the big news of 1928 was Henry Ford's new Model A. Chevrolet responded to the Model A with improved passenger cars that were known as the National Series AB. A "light delivery" commercial chassis was essentially the same as the car unit but with heavier-duty springs. The Series AB light delivery had an updated 171-cid four-cylinder ohv engine rated at 35 bhp. Wood-spoke artillery wheels with demountable rims were standard, but the steel-disk units used on Chevy's cars were available. The company priced the bare light delivery chassis at $375. That figure included front and rear fenders, running boards, and hood. This wood delivery body was made by Hayes Body Corporation. The body design includes a single door on each side, and two panel-type side-hinged doors at the rear. It has removable wood side panels and a rear jump seat that allow the panel-type truck to be converted into a "woody" station wagon.

1928 PICKUP

This meticulously restored 1928 ½-ton National AB pickup wears a body crafted to match the ones produced by the York-Hoover Body Corporation of Pennsylvania. Though the external facing of the body is .040-inch steel, the underlying structure is all wood. The sliding rear window, spare tire mounted at the back of the cab, and integrated bed/cab layout are all unique features of the York-Hoover body design. The 35-hp four-banger could push the pickup to a top speed of about 30 mph. Note the prominent fan shroud, one bit of evidence that the new longer wheelbase chassis was designed to accept the "Stovebolt-Six" engine that would debut for 1929.

The Lowest Ton-Mile Cost provided by Chevrolet Trucks at *Amazing Low Prices*

for Economical Transportation
CHEVROLET

LIGHT DELIVERY **$375** (Chassis Only) f. o. b. Flint, Mich.

UTILITY TRUCK **$495** (Chassis Only) f. o. b. Flint, Mich.

By removing the rear deck and installing an inexpensive slip-on box, the Chevrolet Roadster can easily be converted into a speedy and economical light delivery unit that is ideal for marketing, hauling feed, transporting tools, etc. The slip-on box can be supplied by any Chevrolet dealer.

The ROADSTER **$495** (Box extra) f. o. b. Flint, Mich.

Powered by a valve-in-head motor that is famous the world over for its amazing endurance and efficiency...built with a margin of over-strength in every unit...and incorporating the most advanced engineering design throughout—Chevrolet trucks have repeatedly demonstrated their ability to deliver the world's lowest ton-mile cost!

For Every Line of Business

This matchless economy has been experienced by users in every line of business and under every condition of road and load. Merchants, manufacturers, contractors, farmers and many other users have learned that no other haulage unit does its job so satisfactorily at such low cost.

The Chevrolet Motor Company has always believed that low ton-mile cost is a fundamental requirement in commercial transportation—and every advancement in the design and construction of Chevrolet trucks has been made in the interest of economy...to provide longer life, greater dependability and more efficient operation.

Rugged Construction

Inspect a Chevrolet truck chassis—and you will instantly see how adherence to this principle has influenced Chevrolet design. The banjo-type rear axle is big and sturdy—built to stand up indefinitely under every condition of usage. The frame is constructed of heavy channel steel, rigidly reenforced to withstand the twisting and weaving strains of rough roads. And the long semi-elliptic springs are extra-leaved to cradle the heaviest loads over every type of highway. Furthermore, the benefits of great volume production and scientific management have been utilized to make the first cost as low as possible—a fact that is strikingly demonstrated in the amazing low prices of today's Chevrolet trucks.

Trial-Load Demonstration

Your Chevrolet dealer can provide a body type designed especially for your business and will gladly arrange a trial-load demonstration. See him today.

CHEVROLET MOTOR COMPANY, DETROIT, MICHIGAN
Division of General Motors Corporation

QUALITY AT LOW COST   QUALITY AT LOW COST

By 1928, the Chevrolet truck light-duty line—now dubbed National AB—had grown to a 107-inch wheelbase. Four-wheel brakes (still mechanical) were now standard; previous models had brakes only at the rear. Wood-spoke wheels were still common on '28 Chevy trucks, but they would soon be phased out entirely in favor of steel disc or wire wheels. Chevrolet enjoyed a particularly successful year in 1928, as Ford production was disrupted by the changeover to the Model A. The ad shown here touted Chevrolet economy and reliability while showcasing the company's commercial offerings.

1929 DUMP TRUCK

In its sales battle with Ford, Chevrolet landed a solid punch in 1929: a cast-iron ohv six-cylinder engine for little more than the cost of comparable four-cylinder Fords. Displacement of the Chevy six was 194 cubic inches, and the new engine's rated horsepower was 46. For comparison, the four used by the bowtie brand the previous year made 35 bhp and Ford's Model A was good for 40 horsepower. The 1½-ton LQ was Chevrolet's brawniest truck. The new straight-six engine was utilized intact for the LQ trucks, but it was backed up with a four-speed transmission, rather than the three-speed unit used in passenger cars and the lightest-duty trucks. The LQ series also benefited from a wheelbase stretched to 131 inches. Chevy commercial vehicles now came with standard disc wheels. In 1929, Chevrolet only sold truck chassis and cabs; specialized bodies were supplied by outside firms. The original owner of this truck purchased a dump box made by Kennedy Manufacturing of Streator, Illinois. A hand-crank-operated hoist lifts the dump box.

1929 DUMP TRUCK | 19

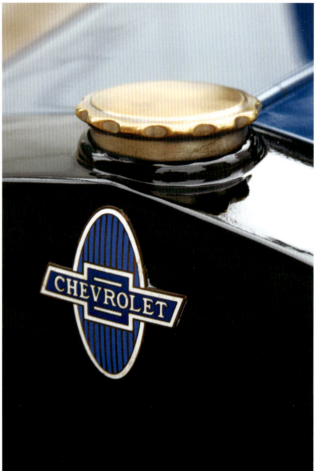

20 | **CHEVY TRUCKS**

1929 DUMP TRUCK | 21

1930 PICKUP

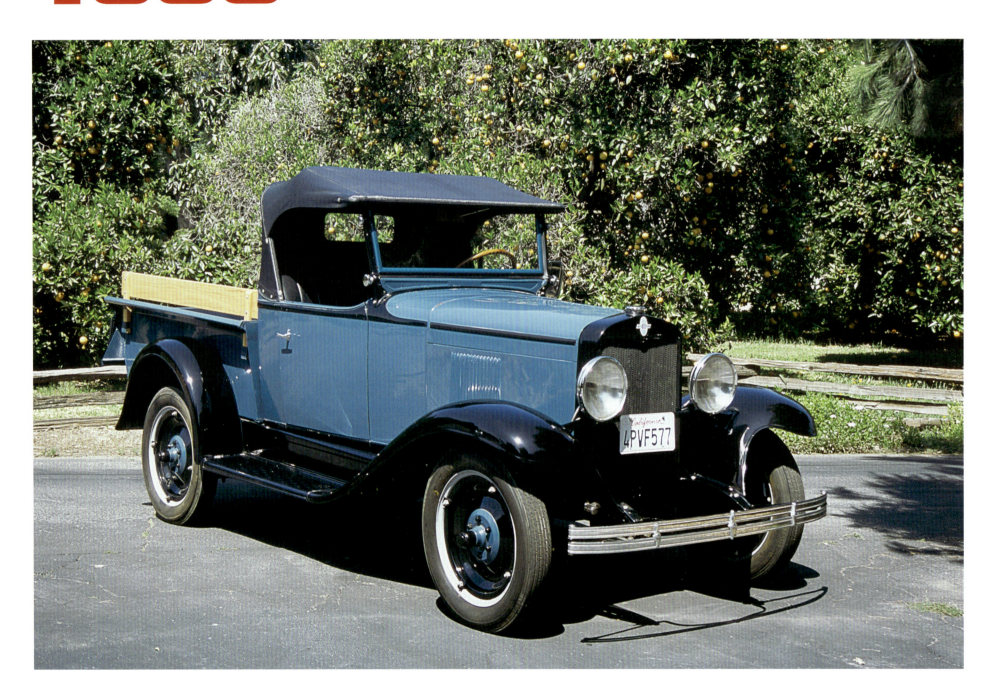

The $400 Series AD roadster delivery was the spunkiest member of Chevrolet's 1930 truck lineup. It was powered by the 194-cid ohv inline six—the "Cast-Iron Wonder"—that Chevrolet had introduced for 1929, though stated horsepower was now up to 50 (from an introductory 46) at 2600 rpm. A new instrument panel featured round, dark-faced dials. Many body parts were shared with Chevrolet's passenger-car roadster, including the hood, fenders, cowl, doors, and a painted version of the passenger cars' grille shell. Pickup beds, however, were made by several manufacturers. Among them was Martin-Parry, which General Motors purchased in late 1930, guaranteeing its own supply of these units in the future. Despite tough economic times and slumping production, Chevrolet built its 7-millionth vehicle this year.

1930 PICKUP

1933 SEDAN DELIVERY

Chevrolet had been steadily improving its six-cylinder engine for greater power and durability. By 1933, the ohv six displaced 207 cid and produced 65 horsepower. The Sedan Delivery bodystyle provided a dignified way for businesses to make light deliveries. The carriage lights were not only an elegant touch, but their red, rear lenses provided an extra degree of visibility at night.

1933 SEDAN DELIVERY | 25

1936 CANOPY DELIVERY

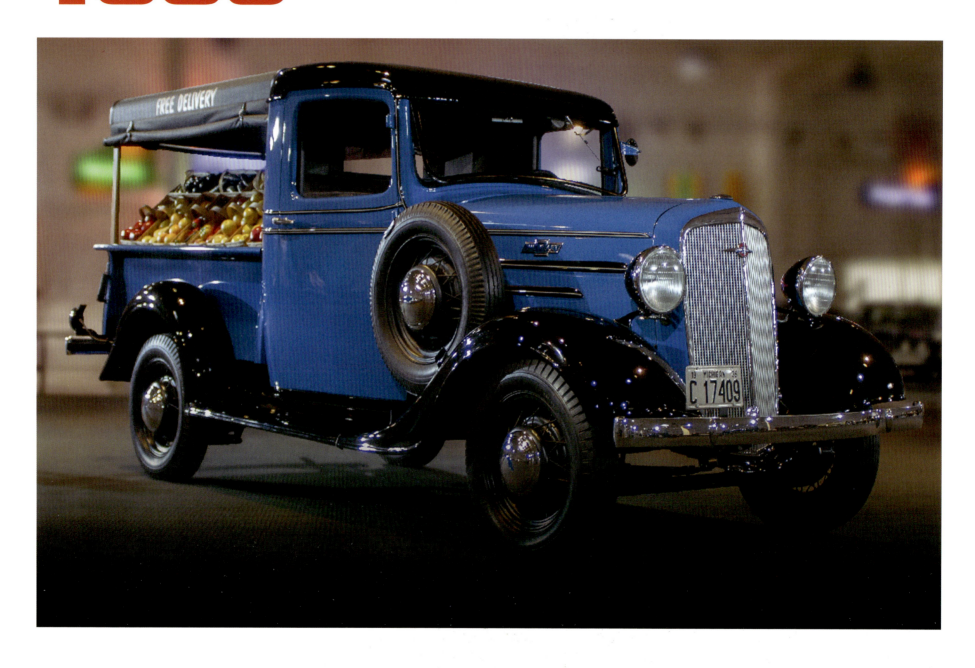

26 | **CHEVY TRUCKS**

"Chevrolet Perfected Hydraulic Brakes" were the big news mechanically for the 1936 Chevrolet lineup; both trucks and cars got them. Chevrolet stuck with mechanical brakes long after most other makes had switched to hydraulics, although rival Ford wouldn't adopt hydraulic brakes until the 1939 model year. This 1936 Chevrolet ½-ton Canopy Delivery wears wire wheels that were being phased out in favor of steel wheels. The Canopy Delivery with open-air sides and roll-down curtains rode a 112-inch wheelbase and started at $577; its steel-sided Panel Delivery sibling started at $12 less. Both had a

1936 CANOPY DELIVERY | **27**

28 | **CHEVY TRUCKS**

115-cubic-foot storage capacity. The 1½-ton Chevrolet trucks still offered plenty of variety; Canopy Delivery (1) and Open Express Pickup (2) models rode a 131-inch wheelbase, while the stakebed model (3) rode a 157-inch span. Chevrolet also offered 1½-ton truck chassis (4) for the tractor-trailer hauling. Tractor trailers were becoming increasingly popular in the mid Thirties.

1936 CANOPY DELIVERY

1936 PICKUP

Midway through the 1936 season, Chevrolet pickups got new cabs with one-piece, all-steel construction and lower, more rounded styling. A single side-mount spare tire was an oft-seen option. The trustworthy Chevrolet six remained at 207 cubic inches, but it was upgraded to produce 79 horsepower, up from 60. Improvements included a new down-draft carburetor, reworked camshaft, and full-length water jackets around the cylinders for improved cooling. The "clear-vison" instrument panel was handsome in its simplicity, as was the rest of the truck. Significantly more-modern styling, with a streamlined, art-deco look borrowed from the passenger car line, was on the way for 1937.

1936 SUBURBAN CARRYALL

The Suburban Carryall wagon debuted for 1935 and was carried over into 1936 with minor changes. At $685, it was the priciest of Chevrolet's ½-ton models. Eight-passenger seating was standard, though accessing the rear seats through the two front doors or the dual rear doors must have been tricky. The Suburban was an innovative vehicle; it was one of the first all-steel station wagons and is sometimes credited as "the first SUV," though that term would not appear for another 50 or so years. All 1936 Chevrolet trucks got new dashboards with a glovebox and relocated instruments (they were now positioned in front of the driver instead of the center of the dash). A mild exterior facelift included a revised grille and hood sides and fenders with slightly "skirted" sides.

1936 SUBURBAN CARRYALL

1941 PICKUP

The 1941 Chevrolet commercial lineup was the broadest in the company's history. There were two engines, three transmissions, five axle ratios, and nine wheelbases. This ½-ton pickup shows that the new design made for a streamlined workhorse. Driver comfort was addressed with features like a crank-open windshield and a comfortable seat that used a latex-bound hair pad on coil springs.

Chevy trucks were restyled for 1941, and marketers placed the new face front and center in this advertisement.

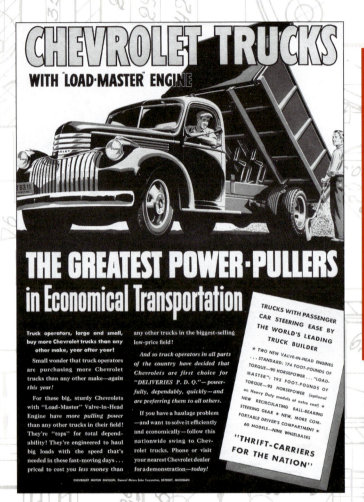

Selling points for the new 1941 trucks included power, economy, dependability, and driver comfort. The copywriters were also sure to point out that Chevy trucks were the best-selling trucks in the business.

(1) Chevrolet's redesigned 1941 trucks also boasted in-fender sealed-beam "safety" headlamps. The standard engine was the same 90-horsepower six used in Chevy's cars. The ½-ton pickup remained the mainstay of the AK series.

(2) The 1941 heavy-duty trucks also used the new styling. This dual-rear-wheel job was primarily meant for tractor-trailer duty.

1942 G-7117 MILITARY

As the war effort progressed, American companies, including Chevrolet, ran patriotically themed magazine ads that were intended to bolster home-front morale with the promise of speedy victory through mass production.

During the war, the American military called on a wide array of manufacturers to provide trucks for the transport of troops and supplies. Chevrolet's specialty was the 1½-ton four-wheel drive G-7100 series. The 1942 Chevrolet G-7117 cargo truck included a front-mounted winch, should the four-wheel drive fail to get the Chevy out of a tough spot. The engine was rated at 83 bhp and the truck had a governed top speed of 48 mph. The G-7117 shown here wears U.S. Navy colors.

1942 G-7117 MILITARY

1942 G-7117 MILITARY | **41**

1946 SUBURBAN

42 | **CHEVY TRUCKS**

After the Allied victory in World War II, Chevrolet, like the rest of American industry, converted back to civilian production as quickly as possible. Chevrolet began production of "Interim" trucks on September 1, 1945. These were not considered 1946 model trucks, even though this line was made through April of that year. Chevrolet's official 1946 models did not enter production until May. These "postwar" models closely mirrored the 1942 models in level of equipment and marked the return of chrome trim. Chrome was a valuable commodity for war production and the final prewar cars and trucks had painted, rather than chrome trim. The ½-ton light-duty 3100 series included pickup, panel, canopy, and Suburban models. The 1946 Model 3106 Suburban was the priciest light-duty model, starting at $1283. The Suburban's body was a modified version of the panel delivery. Like the panel delivery, the Suburban had only two doors and would be a two-door wagon until the 1967 model year. The Suburban and Chevrolet's other ½-ton trucks were powered by a 216.5-cid six good for 90 horsepower. Postwar Suburban and truck customers found spartan, but efficient, interiors.

1946 **SUBURBAN**

1946 PICKUP

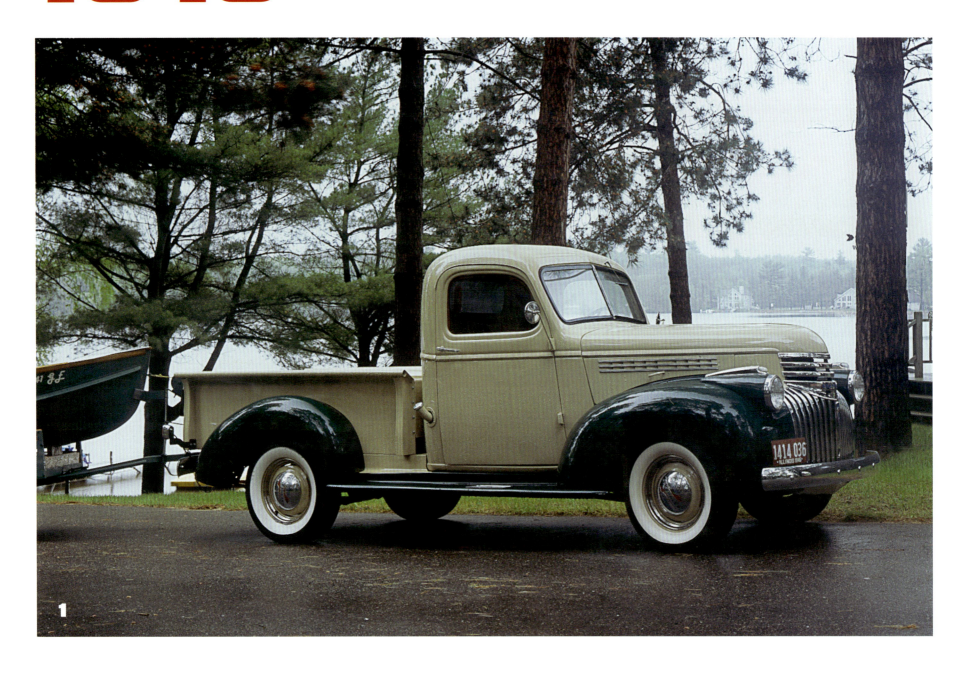

44 | CHEVY TRUCKS

(1) As was common throughout the industry, Chevrolet's 1946 truck line was basically the same as the 1942 offerings. The ½-ton pickup started at $963. (2) The painted grille suggests this pickup was one of the interim models produced in late 1945 and early 1946. After May 1, 1946, chrome grilles were once again standard on Chevy trucks.

1946 PANEL

Chevrolet continued making the 1941-style truck through the end of May 1947. These trucks were titled as '46 or '47 models, depending on when they were originally purchased. Chevrolet sold three basic series of light-duty trucks at this time. The ½-ton models were designated Series 3100, the ¾-ton models were Series 3600, and the one-ton units were Series 3800. This ¾-ton Delivery rode a 125½-inch wheelbase, versus the 115-inch span the ½-ton model used. A one-ton delivery on a 134½-inch wheelbase was also available.

1946 PANEL

1947 PANEL

In mid 1947, Chevrolet, along with corporate cousin GMC, introduced restyled truck lines. These were the first rebodied General Motors postwar vehicles. The so-called "Advance-Design" trucks entered production in May, and officially went on sale on June 28, 1947. This ½-ton Deluxe Panel Delivery is a 1947 model.

1947 PANEL | 49

1948 CANOPY EXPRESS

A variation on the panel truck was the canopy express, which had large display areas cut into its sides and back. This 1948 ½-ton canopy express was joined by a larger one-ton version. Chevy's 1948 trucks were nearly identical to the '47s. Since the Advance-Design trucks were selling briskly in postwar America, there was little incentive, or need, for change.

1948 CANOPY EXPRESS

A 1948 brochure touted the complete line of Chevrolet Advance-Design trucks, that according to the company were "The Choice of the Nation" and "The Leaders in Truck Value."

The new Dubl-Duti forward-control chassis was said to be suited to 9- or 10-foot custom-built panel bodies that could provide double the cubic load space of a conventional panel truck. Chevy offered light-duty Thriftmaster, heavy-duty Loadmaster, and cab-over-engine models. A wide variety of pickups, panels, chassis-cabs, school-bus chassis, and stake trucks were offered. Stake trucks were available in 17 models with payload capacities ranging from 1900 to 10,500 pounds. Promotional materials touted the Advance-Design trucks' combination of power and thrift. Chevy ad writers said the division's new trucks provided the three essentials for successful operation: low first cost, low operating cost, and low maintenance cost.

1949 PICKUP

Chevrolet trucks for 1949 looked the same, but a series of engineering improvements made them stronger and more solidly built. Changes included the way the cab was mounted to the frame, additional bracing for the radiator support and hood, and a brighter taillight. Also, each truck's series was now identified by four-digit chrome numbers on both sides of the hood. Trucks in the 3100 (½-ton) to 4000 (1-ton) series used the 216.5-cid 90-bhp Thriftmaster Six. The 5000 and 6000 (1½ ton) series received a standard 235-cid Loadmaster six, which offered little improvement in horsepower but usefully more torque. Chevy offered the bigger six as an option on the 4000 series trucks.

1949 PICKUP | 55

1949 TOW TRUCK

This 1949 4400 (1½-ton) tow truck wrecker rides a 161-inch wheelbase and carries the standard four-speed manual transmission. This truck is fitted with a Holmes 515E wrecker body.

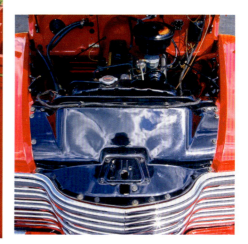

1949 TOW TRUCK | 57

1950 SEDAN DELIVERY

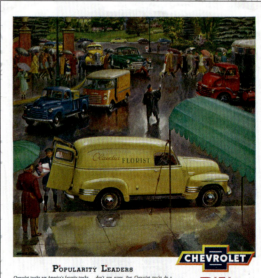

Chevrolet trucks were well-liked by businesses, as this May 1950 magazine ad suggests. Optional chrome trim was late-Art Deco.

The 1950 Sedan Delivery production for the model year topped 23,000, a record for the body style. The Sedan Delivery was popular with small businesses, as it could handle loads as lengthy as 73 inches; total load space was 92.5 cubic feet. Whitewall tires, which brought a touch of elegance, were extra-cost options. Note the slab-sided front fenders and side-hinged cargo door.

1950 SEDAN DELIVERY | 59

1952 PICKUP

60 | **CHEVY TRUCKS**

Changes to Chevrolet's 1952 truck line were modest, but as the Korean War ground on into 1951, one casualty for '52 was chrome, a material with wartime applications. Chevrolet trucks now had less visual sparkle than the year before. The change was most noticeable on the grillework, which looks sober and workman-like on this two-ton '52 pickup. Note the side-mounted spare tire and sun shields for the side glass. Lever-style door handles were replaced by a push-button design this year, and visibility while backing up was improved by a greater curve in the cab's quarter windows. Chevrolet offered two engines in light-duty models: the familiar 216.5-cid six that produced 92 bhp and a 235.5-cid six with 105 horsepower. The base pickup cost $1407 and had a load capacity of 1680 pounds.

1952 PICKUP | 61

1954 PICKUP

The split windshield was replaced by a one-piece curved windshield for 1954, as seen on the 3100 Series pickup and the hefty 3800 Series that rode a 137-inch wheelbase and could be fitted with a variety of cargo beds. The buyer of the 3100 pickup on this page added the available spotlight, side-mounted spare, sun visor, and DeLuxe hood ornament. Chrome (on a new, split grille) was back in '54. The 216.5-cid "Thriftmaster" six was rated at 112 horsepower. Gauges were redesigned on the 3100 Series pickups.

1954 PICKUP | 63

1955 SEDAN DELIVERY

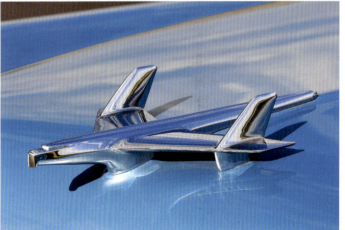

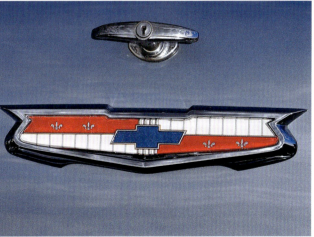

Part car and part truck, the sedan delivery had been on the American motoring scene since the late Twenties. However, by the mid Fifties, the sedan delivery was entering into a decline pushed along by a growing market for pickup trucks that were getting better looking and more comfortable. Given that, the 1955 Chevrolet sedan delivery seen here is quite a rarity with just 8135 built. The 1955 sedan delivery shared the car line's sleek new styling and the option of a sensational new V-8 engine. In terms of profile and dimensions, the sedan delivery matched up with Chevy's new Handyman two-door station wagons. They were 197.1 inches long and stood 62.1 inches tall unloaded. The sedan delivery's plywood load floor was 81.8 inches long and 62 inches across at its widest point. With an internal height of 36 inches, cargo capacity was approximately 91 cubic feet.

1955 SEDAN DELIVERY | 65

This sedan delivery has the standard 235.5-cid inline six-cylinder engine, rated at 123 bhp at 3800 rpm, and three-speed stickshift. A 265-cid ohv V-8 with 145 horsepower was optional. Base price for the sedan delivery was $1699. Factory options on this sedan delivery include a passenger seat, heater, tinted windshield, oil filter, and whitewall tires.

1955 **SEDAN DELIVERY** | **67**

1955 PANEL

For 1955, "first series" Chevrolet trucks with Advance-Design cues (such as this 3100 Panel) were available for just three months. New styling arrived later in the model year with the second series.

Advance-Design was still hyped in this truck-family ad from January 1955.

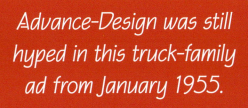

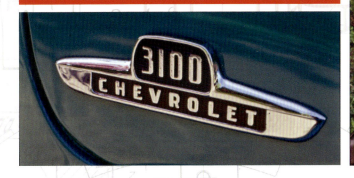

1955 PANEL | 69

1955 PICKUP

The 1955 truck line's "second series" (now available with Chevy's new 265-cid V-8) flaunted startlingly new sheetmetal. The design included integrated front fenders, a dramatically wrapped windshield, hooded headlamps, and oblong grille—variations of what were used on Chevy's restyled 1955 cars. The changes were most dramatic on the Cameo Carrier ½-ton pickup, which featured a fiberglass-fendered straight-sided cargo bed in place of the normal Stepside steel bed. When new, the Cameo's high price made it a slow seller, but over time, it became a much-revered collectible. Another '55 Cameo, in rear view, shows the full-body cargo bed that was made possible by the slab-sided fiberglass fenders. Bombay Ivory with red accents was the only color combination offered. Inside, the Cameo was distinctly carlike, with spiffy upholstery and a modern dash dominated by a hooded, triangular recess that contained the speedometer.

1955 PICKUP | 71

1956 PICKUP

This '56 ½-ton pickup is powered by the 265-cid V-8 that had been introduced by Chevy for 1955. Called "Trademaster" for truck applications, the engine produced 155 hp. As small trucks became increasingly "personal," Chevy creature comforts and accessories were emphasized—such as the whitewalls and full wheel covers seen here. The hood emblem was new for '56, and the side nameplate was shifted to a location above the fender crease. Cabins were functional but hardly less comfortable than what a driver would find in a Chevy car. Although Chevy's uplevel Cameo had a slab-sided cargo bed, the ½-ton carried on with the venerable (and very practical) Stepside design, which would find renewed popularity—as a style statement—four decades later.

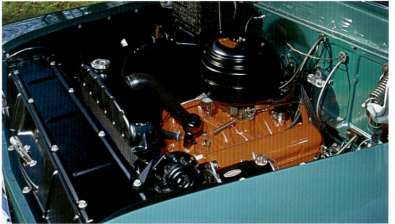

1956 PICKUP | 73

1958 PICKUP

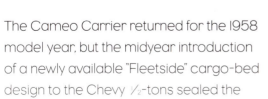

The Cameo Carrier returned for the 1958 model year, but the midyear introduction of a newly available "Fleetside" cargo-bed design to the Chevy ½-tons sealed the fate of the slow-selling Cameo. The new full-width steel-bed Fleetside looked much like the Cameo and cost just $1900 versus Cameo's $2231. Plus, the Cameo's narrow fiberglass bed held less cargo than the Fleetside's bed. Cameo production was discontinued after just 1405 of the '58 models had been built.

1958 PICKUP | 75

1958 PICKUP

Chevy's light trucks adopted dual headlights and the "Apache" name for 1958. February 1958 brought a newly available full-width "Fleetside" cargo bed to the Chevy pickups.

1958 PICKUP | 77

The 1958 Fleetside had 50 percent more cargo capacity than the narrow-box bed of the traditional "Stepside" pickups, but it cost just $16 more. Interiors were essentially carried over from previous years and featured a fuel tank behind the seat. Most '58 Chevy pickups were powered by an 145-bhp 235-cid six-cylinder engine. A 283-cid V-8 with 160 horsepower was optional.

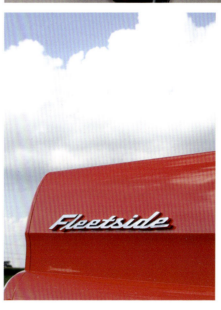

1958 PICKUP | 79

1959 PICKUP

Chevrolet introduced its first four-wheel drive pickups in mid 1957. NAPCO—for Northwestern Auto Parts Company of Minneapolis—supplied a 4WD conversion kit that was installed at the factory. The 4WD conversion added around 50 percent to the price of a Chevrolet truck, and few early Chevy 4WD trucks were sold. This 1959 Chevrolet 3100 Apache Fleetside pickup

was built during the last year that Chevrolet used the NAPCO system. Chevy light-duty trucks were redesigned for 1960; they traded their beam front axle for independent front suspension. The NAPCO gear no longer fit, so Chevrolet engineered its own four-wheel drive. The 4WD option required the 135-bhp "Thriftmaster" six and four-speed manual transmission. The system included a two-speed transfer case with a low range for off-road crawling. It would take several decades before 4WD really caught on with pickup buyers.

1959 PICKUP | 83

1959 EL CAMINO

Chevy countered Ford's Ranchero in 1959 with the graceful El Camino, which shared its "bat wing" styling with the car line. Its six-foot-long bed could carry up to 1150 pounds.

1959 EL CAMINO | **85**

1960 PICKUP

Chevy's 1960 trucks were fully redesigned. Cabs were lower, roomier, more durable, and arguably more contemporary in appearance, even though the double nostrils above the grille looked like a throwback to the company's 1959 passenger car styling. Chevy's light-duty trucks also featured a new chassis. In the front, there was an independent torsion-bar suspension. Out back, the solid rear axle was located by two trailing arms and a lateral stabilizer bar and suspended by two coil springs. The new cab was up to seven inches lower, yet the company said it offered more hip, leg, and head room than the 1959 Chevrolet cab. The new instrument panel placed all gauges and controls directly in front of the driver, and the seat was five inches wider than before. Chevy Stepside pickups used a bed with a wood floor and a grain-tight tailgate. Running boards between the cab and rear fenders assisted side loading. The Apache 10 Stepside was available in short- and long-wheelbase models. The long-wheelbase truck shown here rode a 127-inch span.

1961 CORVAIR RAMPSIDE

Chevy added a line of compact Corvair-based trucks in 1961, which included cargo and window vans along with a pickup. The last was offered as a Rampside that had a fold-down side ramp. The color combination and hubcaps on this truck are attractive, but incorrect.

1961 CORVAIR RAMPSIDE | **89**

1961 PICKUP

The standard Apache pickups received only detail changes for 1961. This Apache 10 pickup has the extra cargo volume of the Fleetside bed. The bed's hardwood floor has steel skid strips to ease sliding of cargo. The standard engine was the 235-cid Thriftmaster six with 135 horsepower. Options on this truck include Custom Comfort seat with six inches of foam rubber, AM radio, wide rear window, two-tone paint, and pointed side moldings.

1961 PANEL

This 1961 panel truck has factory four-wheel drive. The rare 4WD panel truck had a base price of $2985, while the 2WD version cost only $2308. The 2WD truck's torsion-bar independent front suspension was replaced by a solid front axle with leaf springs on 4WD models.

1961 PANEL | 93

1962 PANEL

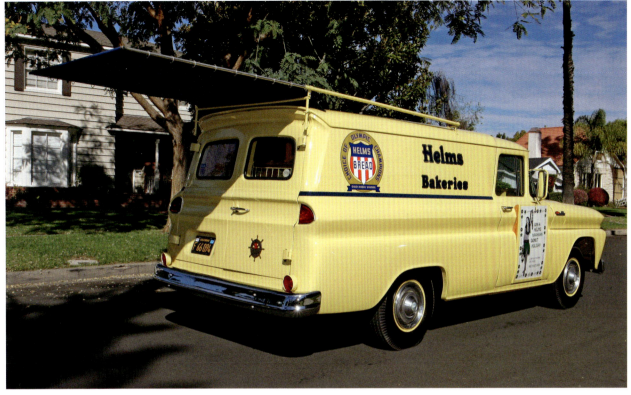

Chevy trucks rolled into 1962 with a modified look that included a hood without the jet-pod styling and two-headlight illumination that replaced the previous quads. This restored Series 10 panel once served a California bakery and incorporated many purpose-built features.

1962 **PANEL** | 95

1963 CORVAIR RAMPSIDE

The Corvair 95 Rampside pickup soldiered on into 1963 largely unchanged. Most people who remember them have a fondness for these Corvair workhorses. Compared to contemporary Volkswagen trucks, they were larger and faster—and because of their superior heater, warmer in winter. In the end, the Corvair trucks were no more successful against Ford's Econoline than the Corvair had been against the Falcon. One by one they disappeared: the Loadside pickup in early 1962, Corvan and Rampside in 1964, and the Greenbrier van in 1965.

1963 CORVAIR RAMPSIDE | **97**

1965 EL CAMINO

Chevy's El Camino resurfaced for 1964, but it was now based on the midsize Chevelle. The 1965 model shown here sported a newly styled front end and revised trim. A base '65 El Camino started at $2270 with a 120-horse 194-cid six.

> The 1964 Chevrolet pickups had revised styling. The changes centered around eliminating the wraparound portions of the windshield (which were outdated by the mid sixties), and included new front pillars.

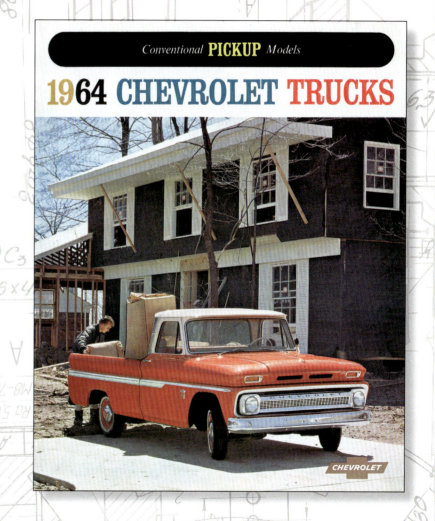

1965 EL CAMINO | 99

1965 PICKUP

Chevy pickups continued to evolve through the 1960s with minor trim and equipment changes each year, plus the occasional new mechanical or convenience feature. This ½-ton 1965 Fleetside used the longer 127-inch wheelbase to accommodate an eight-foot cargo box. The 115-inch short chassis mounted a 6.5-foot bed. Not many folks bought pickups as an alternative to the car in 1965, and Chevy's truck cabs reflected that with a practical but spartan work-oriented cabin decor. The standard 230-cubic-inch inline six is found under this truck's Cardinal Red hood. The 140-horsepower engine mates to a three-speed manual transmission. So equipped, this Fleetside started at $2060: A Stepside model was $16 cheaper. Despite the few changes, Chevrolet truck production topped 600,000 units for 1965—a new record.

1966 EL CAMINO

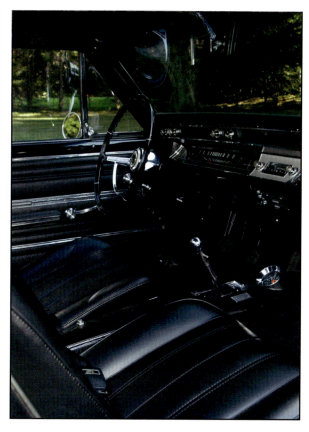

For 1966, the El Camino adopted the midsize Chevelle's attractive new styling and redesigned dashboard.

1966 EL CAMINO | 103

El Camino shared engines with the Chevy's car line, not with trucks. As a result, El Camino was the only light-duty 1966 Chevrolet truck available with the 396-cid "Big Block" V-8 that had horsepower ratings from 325 to 375. There was also a 275-bhp 327-cid V-8. Two sixes were offered: a 194-cid with 120 horsepower and a 230-cid with 140 horsepower. This example has the trusty 283-cid V-8 with 220 horsepower, along with optional bucket seats, floor console, and four-speed manual transmission. El Camino sales were up for 1966 with model-year production topping 35,000.

1966 PICKUP

Again showing little evident year-to-year change was the 1966 edition of Chevrolet's workhorse C-10 Stepside pickup, here in short-box 115-inch wheelbase form. The most obvious visual difference was the new model badge on each front fender. There was news under the hood, since pickups were newly available with a 250-cid inline six and a 327-cid V-8. The latter was offered alongside the 283 V-8, and although the 220-horsepower rating was the same as the smaller engine, the 327 produced more torque. No matter the body color, the interior sheetmetal was painted medium fawn.

1968 PICKUP

Campers were increasingly popular in the late sixties and Chevy offered 8½-foot beds to accomodate the biggest versions. Chevrolet had a new grille for 1969.

After new styling for 1967, the basic design carried over for 1968 with only minor changes to the trim and addition of side-marker lights. Chevrolet claimed to have better rust proofing, particularly for the bed. The new base V-8 was a 307-cid unit with 200 horsepower. A 240-bhp, 327-cid V-8 and a 325-bhp, 396-cid V-8 were also available. For the economy-minded buyer, there were two sixes: a 155-bhp, 250-cid and a 170-bhp, 292-cid.

1970 PICKUP

A slightly revised grille texture with 12 sets of horizontal ribs was one of the subtle changes to the 1970 Chevrolet light-duty trucks. Top-of-the-line trim level was the CST/10 (CST stood for Custom Sport Truck, 10 for the ½-ton series). The upper belt molding seen on this example was optional, but all CST/10s got a wide lower-body molding and bright trim for the pedals. Bucket seats, center console, and tachometer were optional. Engine choices for Chevrolet light-duty pickup trucks ranged from an 155-bhp six to a 310-bhp, 400-cid V-8.

1970 EL CAMINO

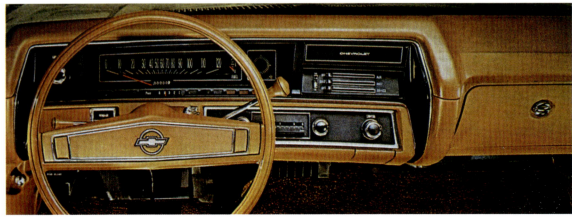

In the original sales brochure, Chevy said there was nothing like an El Camino. Period. It was true that El Camino had few peers. For 1970, the car-based pickup was redesigned up front to closely mimic its Chevelle parent. The SS 396 option added a "big block" V-8 engine, special hood, and sport wheels. A revised instrument panel was also new. A '70 El Camino SS 396 could be fitted with Strato-bucket seats, floor console, and a special instrument package with round gauges.

An ad showcases the large and "small" of the 1970 "Chevy Movers" truck lineup. The heavy-duty Titan 90 tilt-cab semi truck was the new king of the hill. A Fleetside pickup and the off-road ready Blazer carried on.

1970 EL CAMINO | 111

1971 PICKUP

Cheyenne was the new topline interior option for 1971. Cheyenne boasted full carpeting, color-keyed headliner, and a full-depth foam bench seat. Also new for 1971 were standard front disc brakes that offered better fade resistance than drum brakes. The cargo-bed handrail on this truck was a dealer-installed accessory.

1972 PICKUP

For 1972, ½-ton pickups were little changed and carried over the eggcrate grille that debuted for 1971.

114 | CHEVY TRUCKS

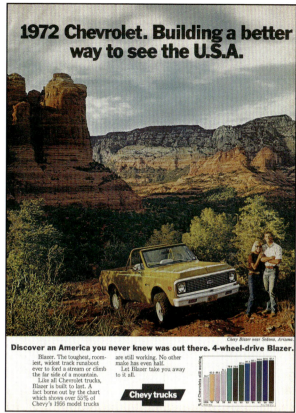

Like its pickup counterpart, the Blazer sport utility vehicle returned for 1972 with few changes.

1972 BLAZER

An answer to the Jeep CJ and Ford Bronco, the Chevy Blazer arrived for 1969 with four-wheel drive only. It continued for 1970 with few changes, plus a new two-wheel drive version. Blazer had a new grille for 1972. A bolt-on removable fiberglass top with integral locking liftgate allowed the Blazer to become a "roadster pickup" or, with the optional rear bench seat, an open-air fun machine with seating for five. The roll bar was after-market accessory.

1972 BLAZER

1975 PICKUP

Chevrolet trucks were redesigned for 1973 with new styling, sturdier frame, and longer wheelbase. The fuel tank was moved from behind the seat to a position on the frame. This eliminated fuel-slosh noise and fuel odor in the cab. For 1975, most Chevrolet light-duty trucks had a catalytic converter to help cope with increasing emissions standards. Chevy continued to offer Fleetside and Stepside beds. This 1975 C-10 has the Stepside bed and base Custom Deluxe interior trim. Engine choices were a 250-cid six, a 350-cid V-8, and a 454-cid V-8. Chevrolet had the best-selling truck line for 1975 with almost 750,000 sold.

1975 PICKUP | 119

1977 EL CAMINO

El Caminos were carried over with little change for 1977; this car-pickup's basic design dated back to 1973. The stacked quad headlights and vertical grille bars on this example identify it as an uplevel Classic model. Lesser versions had single round headlights and a rectangular grille pattern. El Camino Classics started at $4403 and weighed in at 3763 pounds. A 145-hp two-barrel 305 was the base V-8; a 170-hp four-barrel 350 was a $210 option. Eye-catching two-tone paint and sporty rally wheels were also optional.

1977 EL CAMINO

1982 S-10

122 | **CHEVY TRUCKS**

For 1982, Chevrolet replaced its Isuzu-built Luv compact pickup with a domestically produced S-10. The S-10 was larger than the Luv and came in 108- and 118-inch wheelbase versions. A 2.0-liter four-cylinder with 83 horsepower was standard, but a 110-bhp, 2.8-liter V-6 was offered at extra cost.

1982 S-10 | **123**

1983 EL CAMINO

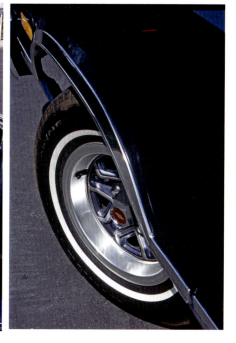

El Camino was redesigned in 1978 and had few changes in the subsequent years. For 1983, El Camino carried on with the four-headlamp appearance that was new the previous year. By this time, it (along with its GMC Caballero twin) was America's only car-based pickup. Prices started at $8191.

1983 EL CAMINO | 125

1988 PICKUP

Chevrolet finally refreshed the full-size pickup line for 1988. Rear antilock brakes were newly available. The Adobe Gold truck above is a K2500 with K designating four-wheel drive and 2500 indicating ¾-ton. It is shown in midlevel Scottsdale trim. The Brandywine Red C1500 truck is a two-wheel drive, ½ ton in topline Silverado trim. Silverados had unique grille and headlamp treatment, choice of cloth or vinyl upholstery, and optional bucket seats.

1988 PICKUP | **127**

1990 PICKUP

For 1990, Chevrolet released this hot 454 SS version of its popular full-size pickup. Exclusively wearing Onyx Black paint, all 454 SS trucks were regular-cab, short-box Fleetsides with a big 230-horsepower 454-cubic-inch V-8 under the hood.

1990 PICKUP | 129

2002 AVALANCHE

For 2002, Chevy introduced a pickup version of its Suburban SUV called the Avalanche. It was similar to a crew-cab pickup, except that the bed was integral with the body, and the two were separated by a removable "midgate" that allowed the 5.3-foot bed to be expanded to 8.1 feet. Avalanche came with a solid three-piece bed cover and covered bins built into the bed walls.

2003 SSR

The SSR two-seat, hardtop convertible pickup arrived during the 2003 model year with an innovative folding hardtop. Based on a shortened two-wheel drive Chevrolet TrailBlazer SUV chassis, it featured retro-flavored styling inspired by Chevy's 1948-53 trucks. Its sole powertrain was a 300-hp 5.3-liter V-8 and a four-speed automatic transmission. Prices started at $41,370. The SSR was the official pace vehicle of the 2003 Indianapolis 500.

2004 COLORADO

A new compact pickup arrived for 2004. Called the Colorado, it was slightly larger than an S-10 and offered two new engines: a 175-horse 2.8-liter four cylinder, and a 3.5-liter five cylinder, both based on the Chevrolet TrailBlazer's inline six. Regular, extended, and crew-cab models were available.

2007 PICKUP

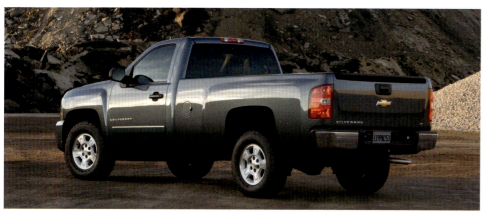

Chevrolet full-size pickups were redesigned for 1999 and Chevrolet shelved the C/K badge in favor of Silverado, formerly its top-level trim. Silverado was completely redesigned again for 2007. Extended-, crew-, and standard-cab models were once again available. A 1500 standard cab is shown above left; a 1500 extended cab is above right. A heavy-duty version of the new Silverado also arrived in 2007. The 2500 HD had a standard 353-horsepower 6.0-liter V-8. A 2500 HD LT Crew Cab is shown on the left.

2014 PICKUP

Chevrolet's redesigned 2014 Silverado benefited from refined powertrains that were fuel efficient for a full-sized pickup. Engine choices were: 285-bhp 4.3-liter V-6, 355-bhp 5.3-liter V-8, and 420-bhp 6.2-liter V-8. As pickups became a common substitute for cars, the interiors became more luxurious. High Country was Silverado's new top trim line with leather upholstery, heated/cooled front seats, and a Bose sound system.

2014 PICKUP | **137**

2015 COLORADO

After a two-year hiatus, Chevrolet's compact pickup truck returned as an all-new model. Both the Chevrolet Colorado and its upscale cousin, the GMC Canyon, were really more midsize than compact. Both a 200-bhp 2.5-liter 4-cylinder and a 305-bhp 3.5-liter V6 were available. Only extended-cab and crew-cab body styles were available. A standard cab was not offered.

2015 COLORADO | **139**

2018 COLORADO ZR2

For 2018, Chevrolet introduced an extreme off-road capable ZR2 version of Colorado. The ZR2 added wider track, raised suspension, 31-inch off-road tires, and Multimatic shock absorbers. It was the only Colorado available with full-time four-wheel drive. A bed-mounted spare tire was optional.

2018 COLORADO ZR2

2019 PICKUP

142 | CHEVY TRUCKS

Silverado was redesigned for 2019. The all-new Silverado was lighter, longer, and taller than the previous model and offered new technology and convenience features. The V-8 engines featured fuel-saving start/stop technology and a Dynamic Fuel Management system that can shut off from two to six cylinders in a variety of combinations to optimize fuel efficiency.

2019 PICKUP | **143**